Bibliographic information published by the German National Library:

The German National Library lists this publication in the National Bibliography;
detailed bibliographic data are available on the Internet at http://dnb.dnb.de .

Imprint:

Copyright © 2017 GRIN Verlag, Open Publishing GmbH
Print and binding: Books on Demand GmbH, Norderstedt Germany
ISBN: 9783668612235

This book at GRIN:

https://www.grin.com/document/385513

Valentina Ocloo

Rectangle-Visibility Representation of Products of Graphs

GRIN Publishing

GRIN - Your knowledge has value

Since its foundation in 1998, GRIN has specialized in publishing academic texts by students, college teachers and other academics as e-book and printed book. The website www.grin.com is an ideal platform for presenting term papers, final papers, scientific essays, dissertations and specialist books.

Visit us on the internet:

http://www.grin.com/

http://www.facebook.com/grincom

http://www.twitter.com/grin_com

Rectangle-Visibility Representations of Products of Graphs

By

Valentina Ekui Ocloo

June 2017

AN ESSAY PRESENTED TO AIMS-GHANA IN PARTIAL FULFILMENT OF THE REQUIREMENTS FOR THE AWARD OF A MASTER OF SCIENCE IN MATHEMATICAL SCIENCES

ACKNOWLEDGEMENTS

Firstly, I would like to express my sincere gratitude to the Almighty God for helping me throughout my studies. I acknowledge the support and cooperation of my parents and siblings, Priscilla Ocloo and Monaliza Ocloo.

I owe my special gratitude to Prof. Neil Turok (Founder of AIMS), Prof. Francis Allotey (President of AIMS Ghana), Prof. Emmanuel Essel (Academic Director of AIMS Ghana) and Beatrice Kwawu (English Language and Communication/Program Officer) and to all other staff and the AIMS community for granting me the privilege of studies. I also do appreciate MasterCard Foundation for their financial support throughout my studies at AIMS.

My supervisor, Prof. Nancy Ann Neudauer (Pacific University, United states) for the continuous support of my essay, patience, motivation, and immense knowledge. Her guidance helped me in all the time of research and writing of this essay. I could not have imagined having a better advisor and mentor for my essay study. I also thank Lord Kavi and to all tutors, for their insightful comments and encouragement.

I also owe my special thanks to all friends and colleagues in AIMS, particularly to Sunday Taiwo, Wisdom Attipoe and William Affum for their precious support and critics in my research and my stay at AIMS Ghana.

DEDICATION

To my parents Mr. and Mrs. Ocloo, and to my dearest friend Collins Abeka.

Abstract

Visibility representation of a graph is a way of assigning the vertices of a graph to objects in a plane and the edges of the graph representing the positioning of the objects in such a way that they see one another. In this work, we consider representations of products of some classes of graphs as rectangle-visibility graphs (RVGs), i.e, graphs whose vertices are rectangles in the plane and edges are horizontal or vertical visibility. We focus on three types of graph products namely: cartesian, direct and strong products. We also investigate representations of products of some classes of graphs such as path, cycle with path, star with path and complete graphs that are RVGs. Furthermore, we discuss why some complete graphs are not RVGs. The results obtained are established by constructive proofs and yield linear-time layout.

Keywords: Rectangle-visibility graph, cartesian product, direct product and strong product.

Contents

List of Figures

List of Symbols and Abbreviations

RVG..Rectangle-visibility graph.

BVG..Bar-visibility graph.

VLSI..Very-Large-Scale Integration.

$G \times H$..Cartesian product of G and H.

$G\square H$..Direct product of G and H.

$G \boxtimes H$..Strong product of G and H.

P_n..Path graph on n vertices.

C_n..Cycle graph on n vertices.

S_m..Star graph on m vertices.

K_n..Complete graph on n vertices.

1. Introduction

1.1 Background of the Study

The history of graph theory may specifically be traced to 1735 when the Swiss mathematician Leonhard Euler solved the Konigsberg bridge problem [1]. The Konigsberg bridge problem originated in the city of Konigsberg. The city had seven bridges which connected two islands. People in the city always wondered whether there was any way to walk over all the seven bridges only once. In 1736, Euler came out with a solution in terms of representing a set of vertices connected by edges (graph), and hence making the problem much easier to solve. This led to the application of graph theory in studies such as chemistry, electrical engineering, transportation network and many others. Another application of graph theory is found in the study of rectangle-visibility graphs. This representation of graphs have been extensively analysed over the past years, because of its application in the design of VLSI (Very-Large-Scale Integration) chip.

Visibility representation of a graph is by assigning the vertices of a graph to objects in a plane and the objects are placed in such a way that they can see each other. The visibility between these objects correspond to the edges of the graph. There have been many visibility representation studies on how to represent graphs. However, the most common ones are the bar-visibility graphs (BVGs) and rectangle-visibility graphs (RVGs). Bar-visibility graphs (BVGs) are those planar graphs whose vertices can be represented by horizontal line segments with adjacency determined by vertical visibility [2].

On the other hand, rectangle-visibility graphs (RVGs) are graphs whose vertices are represented as rectangles in a plane and edges as horizontal or vertical visibility. An RVG is known to be the union of two planar graphs when the vertical and horizontal visibilities in its layouts are considered, and thus deconstructed into a union of two bar-visibility graphs [3]. Hence, we say such graphs have thickness at most two. Shermer [4] showed that it is NP-complete (nondeterministic polynomial time) to determine if a graph is an RVG, and so it is of interest to determine classes of graphs that are and those that are not RVGs.

The focus of this essay is to investigate representations of products of some classes of graphs as rectangle-visibility graphs.

1.2 Rationale of the Project

In making of computer chips, the Very-Large-Scale Integration is usually used, following a method of setting out and fixing transistors on the surface of a small chip in addition to the wired network between components [5]. Thus in designing chips, visibilities are considered between various electrical parts. Meanwhile visibility graphs are limited in their application to VLSI design as a result of the restriction of edges in the graph corresponding only to objects that see one another. However, they form a beginning point for the consideration of representing a physical circuit network as a graph [5].

For such constructions, gates or other chip components represent horizontal bars or rectangles in the plane, and edges correspond to vertical visibilities between bars, or by horizontal and vertical visibilities between rectangles [6]. These horizontal and vertical visibilities in RVGs form a typical example for two-layer chips with wires running vertically on one layer and horizontally on the other layer [6]. Figure 1.1 is an example of the VLSI component.

Figure 1.1: The VLSI component.

1.3   Overview of the Project

The organization of this essay is as follows. In Chapter 2, we provide a literature review on visibility representations, specifically rectangle-visibility representation. In Chapter 3, we present basic definitions and notations from graph theory and rectangle-visibility graphs. In Chapter 4, we study three types of graph products and investigate representations of products of some classes of graphs as rectangle-visibility graphs, and present our result. We also discuss why some complete graphs are not RVGs. Finally in Chapter 5, we give a conclusion of this work and state a possible extension of the results for future work.

2. Literature Review

In 1976, the first study of rectilinear objects in a two-dimensional plane with vertical and horizontal visibility was done by Garey et al. [7]. They recommended a strategy for testing printed circuit boards for the presence of possible (undesired) short circuits changes for the test minimization problem, into one of finding minimum vertex colourings of certain special graphs, called line-of-sight graphs. Also, they considered certain assumptions on the possible types of short circuits and analysed the structure of such graphs and showed that a well known and efficient algorithm can be used to colour them with a small number of colours.

Tamassia and Tollis [8] studied visibility representations of planar graphs. They considered three types of visibility representations, and gave complete characterizations of classes of graphs that admit them. Tamassia and Tollis showed that a bar-visibility graph is a planar graph but not all planar graphs are bar-visibility graphs. Furthermore, they presented linear time algorithms for testing the existence and constructing visibility representations of planar graphs.

The study of rectangle-visibility graphs (RVGs) just like bar-visibility graphs was inspired by studies in [7]. Bose et al. [9] considered classes of graphs that are RVGs. They established that for $1 \leq k \leq 4$, k-trees are RVGs, and that, any graph that can be decomposed into two caterpillar forests is an RVG. Also, they showed that any graph whose vertices of degree four or more, form a distance-two independent set is an RVG. Lastly, any graph with maximum degree four is an RVG. They also distinguished these graphs as collinear, noncollinear and weak rectangle-visibility graphs.

In 1997, Dean and Hutchinson [2] considered representation of bipartite graphs as rectangle-visibility graphs. They showed that for complete bipartite graphs $p \leq q$, $K_{p,q}$ has a representation with no rectangles having collinear side if and only if $p \leq 2$ or $p = 3$ and $q \leq 4$. More generally, Dean and Hutchinson [2] showed that $K_{p,q}$ is a rectangle-visibility graph if and only if $p \leq 4$ and also that every bipartite rectangle-visibility graph on $n \geq 4$ vertices has at most $4n - 12$ edges. In the course of their findings they showed that, though $K_{5,5}$ plus any edge, and $K_{5,5}$ minus any edge are rectangle-visibility graphs; $K_{5,5}$ by itself is not a rectangle-visibility graph.

Also, Kant et al. [10] studied the area requirement of bar-visibility and rectangle-visibility representations of trees in the plane. They proved asymptotically tight lower and upper bounds on the area of such representations, and provided algorithms that construct representations with asymptotically optimal area in linear time.

In 1998, Dean et al. [11] considered representations of unions and products of trees as rectangle-visibility graphs. In their study, they established that the union of any tree (or forest) with a depth-1 tree is an RVG and that the union of two depth-2 trees and the union of a depth-3 tree with a matching are subgraphs of RVGs. They also showed that the cartesian product of two forests is an RVG.

Soon after, Hutchinson et al. [3] studied thickness-two graphs (graphs whose edge set can be partitioned into two planar subgraphs) and their representations as rectangle-visibility graphs and as doubly linear graphs (graphs that can be drawn in a plane as the union of two straight-edged planar graphs). They proved that complete graph K_n is an RVG if and only if $n \leq 8$. In their

4

paper, they showed that RVGs with n vertices have at most $6n-20$ edges different from thickness two graphs with $6n-12$ edges.

Recently, Peterson [5] proved results on rectangle-visibility numbers of graphs such as trees, complete graphs, complete bipartite graphs and $(1, n)$-hilly graphs which are graphs where there is no path of length 1 between vertices of degree n or more.

In the next Chapter, we present basic definitions and notations in graph theory and rectangle-visibility graphs.

3. Preliminaries

In this chapter, a brief introduction to terms and notations needed throughout this work is given. Most of these definitions are standard terms used in graph theory and rectangle-visibility graphs.

3.1 Definitions

The following definitions can be found in [1, 12].

Definition 3.1.1 (Graph). A *graph* G is a pair of sets (V, E) where V is the set of vertices and E is the set of edges.

An edge denoted as vw connects the vertices v and w. Two vertices v and w of a graph G are adjacent if there is an edge vw connecting them. The vertices v and w are then incident with the edge vw. A *loop* is an edge that connects a vertex to itself. *Parallel or multiple edges* are two or more edges that are incident to the same two vertices. The *degree of a vertex v of G* is the number of edges incident with v and is denoted as $\deg(v)$.

Definition 3.1.2 (Subgraph). A *subgraph* $H = (U, P)$ of a graph $G = (V, E)$ is a graph such that $U \subseteq V$ and $P \subseteq E$.

Definition 3.1.3 (Simple graph). A *simple graph* is a graph with no parallel edges or loops.

Example 3.1.4. Figure 3.1 shows a simple graph and a non-simple graph with multiple edges. The vertex set $\{a, b, d\}$ connected with the edges is a subgraph of the simple graph.

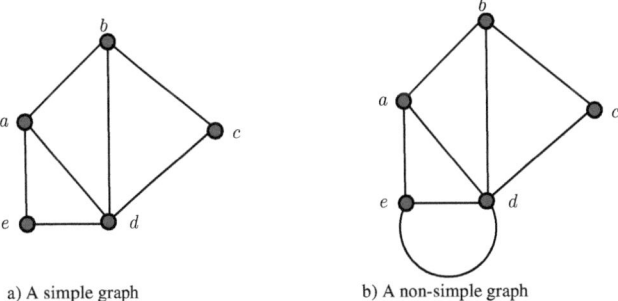

a) A simple graph   b) A non-simple graph

Figure 3.1: A simple graph and a non-simple graph.

Now, we define several specific classes of graphs that we will use in this work.

Definition 3.1.5 (Walk and Path graphs). A *walk* in a graph G is a sequence

$$W = v_0, e_1, v_1, e_2, \ldots, v_{k-1}, e_k, v_k,$$

whose terms alternate between vertices and edges (not necessarily distinct) such that $1 \leq i < k$, the edge e_i has ends v_{i-1} and v_i. A *path* is a walk such that all of the vertices and edges are distinct. We denote a path graph on n vertices by P_n. A typical example of a path graph is shown in Figure 3.1 of the simple graph where the path $P : a \rightarrow b \rightarrow d \rightarrow c$.

Definition 3.1.6 (Cycle graph). A connected graph that is regular of degree 2 is a *cycle graph*. In other words, it consists of a number of vertices connected in a closed chain. A cycle graph with n vertices has n edges and is denoted by C_n.

Definition 3.1.7 (Complete graph). A simple graph where every pair of vertices is connected by an edge is a *complete graph*. We denote the complete graph on n vertices by K_n. A complete graph with n vertices has $n(n-1)/2$ edges.

Definition 3.1.8 (Bipartite graph). A *bipartite graph* is a graph whose vertices can be separated into two disjoint sets U and V in such a way that every edge connects a vertex in U to a vertex in V.

Definition 3.1.9 (Complete bipartite graph). A *complete bipartite graph* is a bipartite graph where every vertex in set U is connected to every vertex in set V. A complete bipartite graph with $|U| = m$ and $|V| = n$ is denoted by $K_{m,n}$.

Definition 3.1.10 (Star graph). A *star graph*, S_n is a connected graph on n vertices where one vertex has degree $n - 1$ and the other $n - 1$ vertices have degree 1. Thus, a star graph is a special case of a complete bipartite graph in which one set has 1 vertex and the other set has $n - 1$ vertices denoted by $S_n = K_{1,n-1}$.

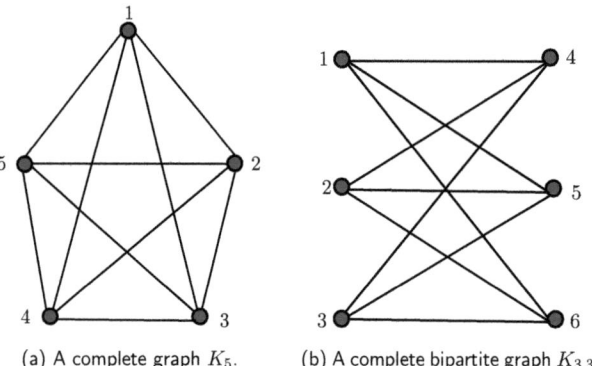

(a) A complete graph K_5.   (b) A complete bipartite graph $K_{3,3}$.

Figure 3.2: Examples of a complete graph and a complete bipartite graph.

In relation to circuit design, there is a need to avoid or reduce crossing of wires since crossing leads to unwanted signals. For that matter, we define graphs that are usually used in avoiding crossing of wires.

Definition 3.1.11 (Planar and plane graphs). A graph G is called *plane graph* if it is drawn in the plane without any edges crossing each other. A graph which can be redrawn in a plane without any edges crossing is called a *planar graph*.

Upon drawing a plane graph it divides the plane into a set of regions known as *faces*, each bounded by edges of the graph. A face can be bounded or unbounded. We call an unbounded face the *outer or infinite face*. We denote the number of faces by f. Examples of infinite faces are f_4 and f_4' shown in Figure 3.3b and Figure 3.3c.

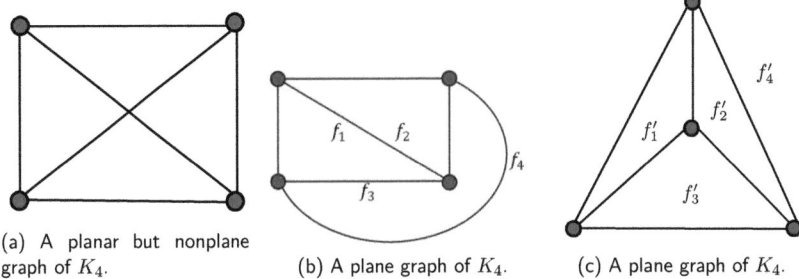

(a) A planar but nonplane graph of K_4.

(b) A plane graph of K_4.

(c) A plane graph of K_4.

Figure 3.3: A plane drawing of K_4.

Theorem 3.1.12 (Euler's formula [12]). *Let G be a planar drawing of a connected planar graph and let n, m and f denote respectively the number of vertices, edges and faces of G. Then, $n - m + f = 2$.*

Let us note that not all graphs are planar. A graph which is not a planar graph is called non-planar. Examples of non-planar graphs are $K_{3,3}$ and K_5.

Definition 3.1.13 (Thickness of a graph). The thickness (t) of a graph (G) is the smallest number of planar subgraphs whose union is G.

Next, we present a theorem on thickness of complete graphs.

Theorem 3.1.14 ([13]). *The thickness of a complete graph on n vertices, K_n is*

$$t(K_n) = \left\lfloor \frac{n+7}{6} \right\rfloor,$$

except when $n = 9, 10$ for which the thickness is 3.

Now, we define types of visibility graphs that are necessary in this work.

Definition 3.1.15 (Bar-visibility graphs (BVGs) [2]). *Bar-visibility graphs* are those planar graphs whose vertices can be represented by horizontal line segments with adjacency determined by vertical visibility. Figure 3.4 illustrates an example of a bar-visibility representation of a complete graph K_4.

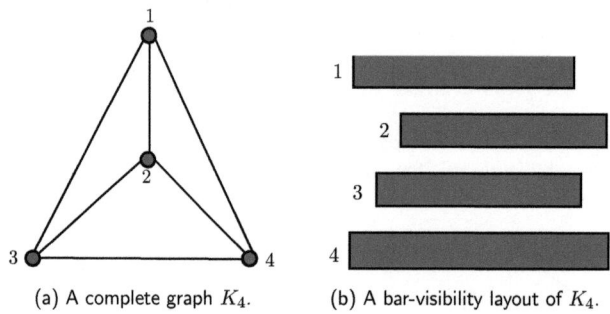

(a) A complete graph K_4.   (b) A bar-visibility layout of K_4.

Figure 3.4: A complete graph K_4 and its bar-visibility representation.

Definition 3.1.16 (Rectangle-visibility graphs (RVGs)[2]). A graph G is a *rectangle-visibility graph* if its vertices can be represented by closed rectangles in the plane with sides parallel to the axes, pairwise disjoint except possibly for overlapping boundaries, in such a way that two vertices u and w are adjacent if and only if each of the corresponding rectangles is vertically or horizontally visible from the other. We call such a layout as rectangle-visibility representation which is denoted as rv-representation and an example is shown in Figure 3.5.

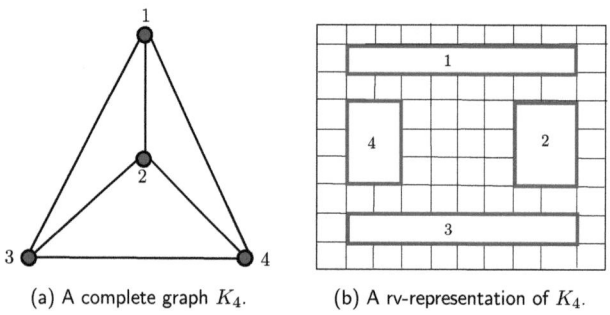

(a) A complete graph K_4.   (b) A rv-representation of K_4.

Figure 3.5: A complete graph K_4 and its rv-representation.

Definition 3.1.17 (Noncollinear rectangle-visibility representation [2]). A rectangle-visibility representation is called *noncollinear* if no two rectangles have collinear sides, that is, no two rectangles have endpoints with the same x or y-coordinate. We call such a layout a noncollinear rv-representation. A typical example is illustrated in Figure 3.6 whiles Figure 3.5b depicts a collinear rectangle-visibility representation.

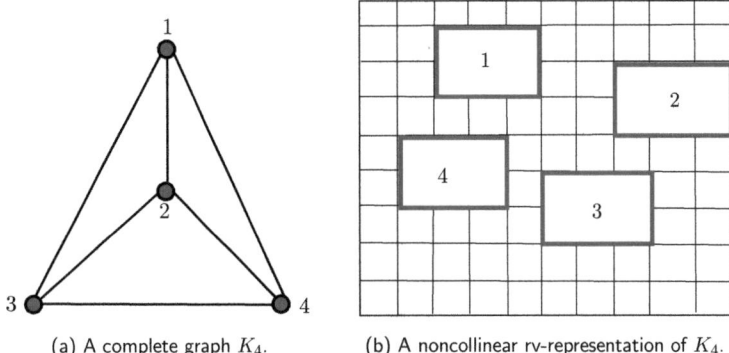

(a) A complete graph K_4. (b) A noncollinear rv-representation of K_4.

Figure 3.6: A complete graph K_4 with its noncollinear rv-representation.

Definition 3.1.18 (Unit rectangle-visibility graph [6]). A graph G is a *unit rectangle-visibility graph or URVG* if its vertices can be represented by closed unit squares in the plane with sides parallel to axes and pairwise disjoint interiors, in such a way that two vertices are adjacent if and only if there is an unobstructed non degenerate (positive width) horizontal or vertical band of visibility joining the two rectangles. Figure 3.7 depicts a unit rectangle-visibility representation of a graph G.

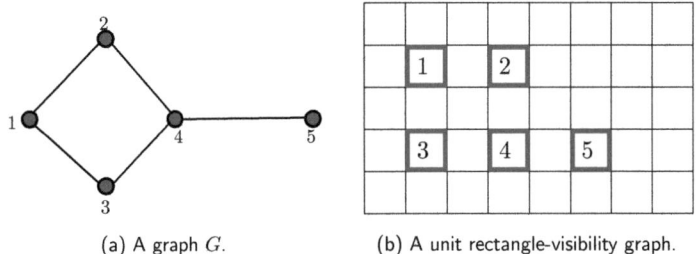

(a) A graph G. (b) A unit rectangle-visibility graph.

Figure 3.7: A graph G with its unit rectangle-visibility representation.

In the next Chapter, we define graph products and focus on three types of graph products. We then investigate representations of products of some classes of graphs as rectangle-visibility graphs.

4. Products of Graphs as Rectangle-Visibility Graphs

In this chapter, we introduce the idea of products of graphs and then focus on three types of graph products namely: cartesian, direct and strong products. We also define some classes of cartesian product graphs and then finally, we investigate representations of products of some classes of graphs as rectangle-visibility graphs.

4.1 Definitions of the types of Graph Products

Most of these definitions of graph products can be found in [14].

Definition 4.1.1. We say a *graph product* is a way of multiplying two graphs to obtain a new graph.

Now let us consider two graphs, say G and H where the set of vertices of G is given by $V(G) = \{a_1, a_2, \ldots, a_i\}$ and the set of vertices of H is given by $V(H) = \{g_1, g_2, \ldots, g_n\}$ for $i, n \in \mathbb{Z}$. Then, the set of vertices $V(G) \times V(H)$ of the product of G and H is

$$\{(a_1, g_1), (a_1, g_2), \ldots, (a_1, g_n), (a_2, g_1), (a_2, g_2), \ldots, (a_2, g_n), \ldots, (a_i, g_1), (a_i, g_2), \ldots, (a_i, g_n)\}.$$

With this idea, we can define the types of graph products as follows.

Definition 4.1.2 (Cartesian product)**.** Given two graphs G and H, their *Cartesian product* $(G \times H)$ is defined as a graph with the vertex set $V(G) \times V(H)$ and any two vertices (a, b) and $(1, 2)$ are adjacent in $G \times H$ if and only if either $a = 1$ and b is adjacent to 2 in H or $b = 2$ and a is adjacent to 1 in G.

Definition 4.1.3 (Direct product)**.** Given two graphs G and H, their *Direct product ($G \square H$)* is defined as a graph with the vertex set $V(G) \times V(H)$ and any two vertices (a, b) and $(1, 2)$ are adjacent in $G \square H$ if and only if a is adjacent to 1 in G and b is adjacent to 2 in H.

Definition 4.1.4 (Strong product)**.** Given two graphs G and H, their *Strong product ($G \boxtimes H$)* is defined as a graph with the vertex set $V(G) \times V(H)$ and any two vertices (a, b) and $(1, 2)$ are adjacent in $G \boxtimes H$ if and only if

- $a = 1$ and b is adjacent to 2, or $b = 2$ and a is adjacent to 1,
- a is adjacent to 1 and b is adjacent to 2.

In other words, the strong product is the combination of the characteristics of both cartesian and direct products. Next, we define some classes of cartesian product graphs that we will use in the next section.

Definition 4.1.5 (Grid or Lattice graph). A *grid graph* is the graph cartesian product, $P_n \times P_m$ of path graphs on m and n vertices. It is also an $m \times n$ lattice graph denoted as $G_{m,n}$.

Definition 4.1.6 (Hypercube graph). A *hypercube graph* Q_n is defined as the graph cartesian product of path graphs $\underbrace{P_2 \times P_2 \times \ldots \times P_2}_{n}$.

Definition 4.1.7 (Stacked prism graph). A *stacked prism graph* $Y_{m,n}$ is a simple graph given by the graph cartesian product $Y_{m,n} = C_m \times P_n$.

Definition 4.1.8 (Book graph). A *book graph* B_m is defined as the graph cartesian product of $S_{m+1} \times P_2$, where S_m is a star graph and P_2 is the path graph on two vertices.

Definition 4.1.9 (Torus grid graph). A *torus grid graph* $T_{m,n}$ is the graph formed from the cartesian product $C_m \times C_n$ of the cycle graphs C_m and C_n.

In the following sections, we investigate representations of cartesian, direct and strong products for path, cycle, star and complete graphs as rectangle-visibility graphs. We abbreviate the vertex (a_i, a_j) as a_i, a_j and use them interchangeably.

4.2   Cartesian Product

In the cartesian product, Dean et al. [11] established on rectangle-visibility representations for products of graphs as given below.

Theorem 4.2.1 ([11]). *If the graphs G and H are both BVGs, then $G \times H$ is an RVG. Analogous statements hold if G and H are noncollinear or weak BVGs.*

With regards to the two graphs being BVGs, Tamassia and Tollis [8] showed that BVGs are planar graphs but not all planar graphs are BVGs. Hence, if G is the cartesian product of two planar graphs, then G has thickness at most 2 which follows that G is an RVG. The following results, though discovered independently, can be stated as corollaries to Theorem 4.2.1 and we give some illustrative examples. We start with cartesian product for path graphs.

Path graph: We construct the cartesian product for $P_3 \times P_2$ and $P_4 \times P_4$. In Figure 4.1 of the construction of $P_3 \times P_2$, the vertex set $V(P_3) \times V(P_2) = \{(a,1), (b,1), (c,1), (a,2), (b,2), (c,2)\}$. Consider vertices $(a,1)$ and $(b,1)$, since $1 = 1$ and a is adjacent to b in the graph (P_3), we put an edge between $(a,1)$ and $(b,1)$. Hence by adding the edges in that manner, we obtain a grid graph as shown in Figure 4.1. Since $(b,1)$ is adjacent to $(a,1)$, $(c,1)$ and $(b,2)$ in the grid graph $(P_3 \times P_2)$, we represent each vertex as a rectangle of equal size and place the rectangle in a way that it is visible by its adjacency. At the end of the arrangement, we obtain an rv-representation in Figure 4.2.

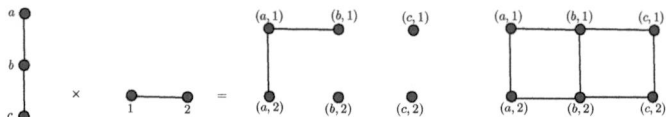

Figure 4.1: Construction of $P_3 \times P_2$.

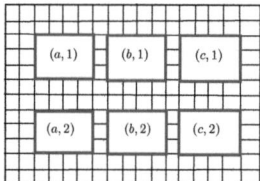

Figure 4.2: A rv-representation of $P_3 \times P_2$.

Next, Figure 4.3 shows the construction of $P_4 \times P_4$ and its rv-representation. From Figure 4.3a, the vertex set $V(P_4) \times V(P_4) = \{(a,1), (b,1), (c,1), (d,1)(a,2), (b,2), (c,2), \ldots, (d,4)\}$. We follow the same approach in generating the earlier graph shown. By that, we obtain a 4×4 grid graph in Figure 4.3a. We follow the earlier procedure in the construction of its rv-representation to obtain Figure 4.3b.

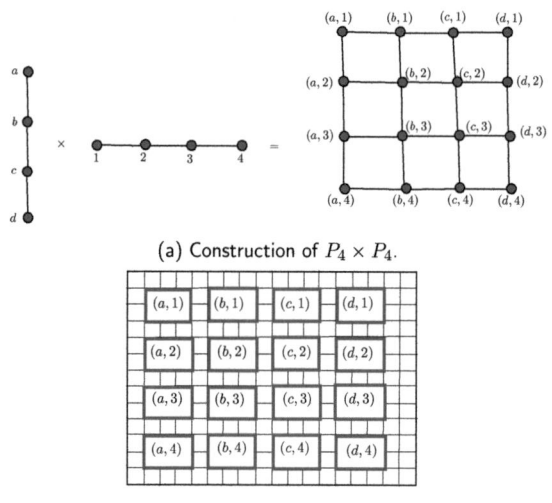

(a) Construction of $P_4 \times P_4$.

(b) A rv-representation of $P_4 \times P_4$.

Figure 4.3: Construction of $P_4 \times P_4$ and its rv-representation.

Corollary 4.2.2. The cartesian product of two path graphs is an RVG. That is, the grid (lattice) graph is an RVG.

Proof. We give the construction.

Suppose P_m is a path graph with m vertices and P_n is a path graph with n vertices. Then, the cartesian product of P_m and P_n gives a grid graph. Now for each vertex of the grid graph obtained, draw a rectangle of equal size. If $m = n$, we obtain $n \times n$ rectangles and if $m \neq n$ we obtain $m \times n$ rectangles. Place each vertex as a rectangle and the adjacency of the vertices representing vertical and horizontal visibilities between the rectangles. By this, we obtain the rv-representation of the cartesian product of P_m and P_n as in Figure 4.4.   □

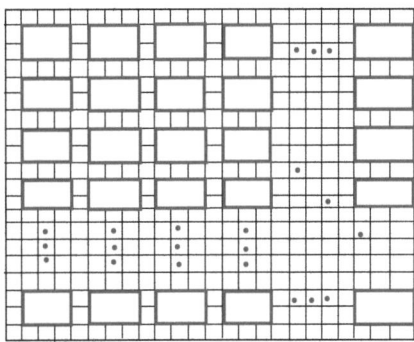

Figure 4.4: A rv-representation of $P_m \times P_n$.

From Figure 4.4, the RVG is collinear since two corners of the rectangles share the same coordinates and can also be represented in noncollinear RVG. Furthermore, since the path graph is a BVG then the cartesian product of the two path graphs are RVGs.

Corollary 4.2.3 ([11]). The hypercube graphs Q_n, for $n = 1, \ldots, 6$, are RVGs.

Q_3 is really a BVG, since it is a 2-connected planar graph. For $Q_4, \ldots, Q_7$, are hypercubes known of thickness only 2. However, it is not known if Q_7 is an RVG. Next, we consider the cartesian product of cycle and path graphs.

Cycle and path graphs: First we construct the cartesian product for C_3 and C_3 and next construct $C_6 \times P_2$ and $C_5 \times P_3$. Figure 4.5 depicts the construction of $C_3 \times C_3$ and its rv-representation. In Figure 4.5a, the vertex set $V(C_3) \times V(C_3) = \{(a, 1), (b, 1), (b, 1), (a, 2), (b, 2), (c, 2)(a, 3), (b, 3), (c, 3)\}$. Without loss of generality, we obtain the torus grid graph and its rv-representation is shown in 4.5b.

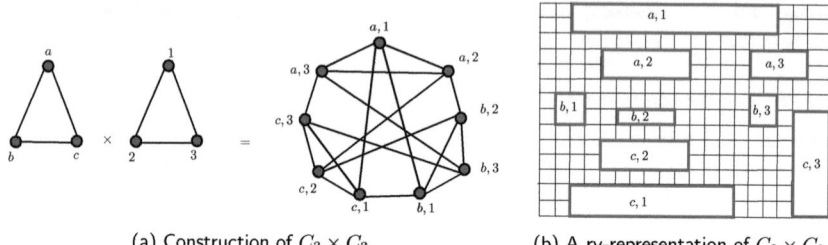

(a) Construction of $C_3 \times C_3$.   (b) A rv-representation of $C_3 \times C_3$.

Figure 4.5: Construction of $C_3 \times C_3$ and its rv-representation.

Now, we investigate the case of cycle with path graphs. Figure 4.6, illustrates the construction of $C_6 \times P_2$ which is a stacked prism graph and its rv-representation. The stacked prism graph obtained from the construction of $C_6 \times P_2$ is shown in Figure 4.6a with its rv-representation shown in Figure 4.6b.

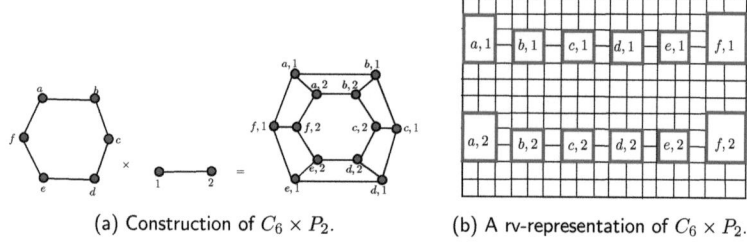

(a) Construction of $C_6 \times P_2$.   (b) A rv-representation of $C_6 \times P_2$.

Figure 4.6: Construction of $C_6 \times P_2$ and its rv-representation.

Similarly, $C_5 \times P_3$ is a rectangle-visibility graph and its construction with the rv-representation is shown in Figure 4.7. In the next corollary, we prove by construction that $C_m \times P_n$ is an RVG.

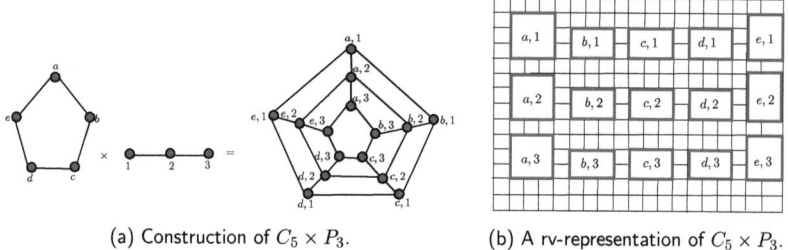

(a) Construction of $C_5 \times P_3$.   (b) A rv-representation of $C_5 \times P_3$.

Figure 4.7: Construction of $C_5 \times P_3$ and its rv-representation.

Corollary 4.2.4. The cartesian product of C_m and P_n is an RVG. That is, the stacked prism graph is an RVG.

Proof. We give the construction.

Suppose that C_m is a cycle graph with m vertices and P_n is a path graph of n vertices. Then, the cartesian product of C_m and P_n gives a stacked prism graph. The stacked prism graph has a circular pattern which is drawn in the following layout. Without loss of generality, we represent each vertex as a rectangle. We let $V_{i,j}$ be the vertex for each $i \in \{1, m\}$ and for each $j \in \{1, n\}$, then there exist n unique cycles in the stacked prism graph of length m and also n unique path graphs respectively.

- First represent $V_{i,1}, i \in \{1, m\}$ forming the cycle graph.

- Next, represent the cycle graph $V_{i,2}, i \in \{1, m\}$ by placing the vertices directly below the previous vertices. That is, $V_{m,2}$ sees $V_{m,1}$ if and only if $m = m$.

- Lastly, repeat the process for the remaining cycles $V_{i,j}$ (i.e. for $j = 3, \ldots, n$) ensuring that all the n paths are also duely represented.

At the end of the layout, we obtain a horizontal and vertical visibilities in rows and columns. The construction of this representation is shown in Figure 4.8.   □

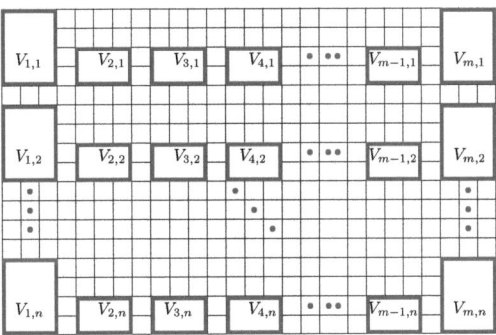

Figure 4.8: A rv-representation of $C_m \times P_n$.

Next, we consider the cartesian product of star with path graphs and complete graphs.

Star with path graphs: Figure 4.9a represents when $m = 7$, we obtain B_7 ($S_8 \times P_2$) and its rv-representation is shown in Figure 4.9b.

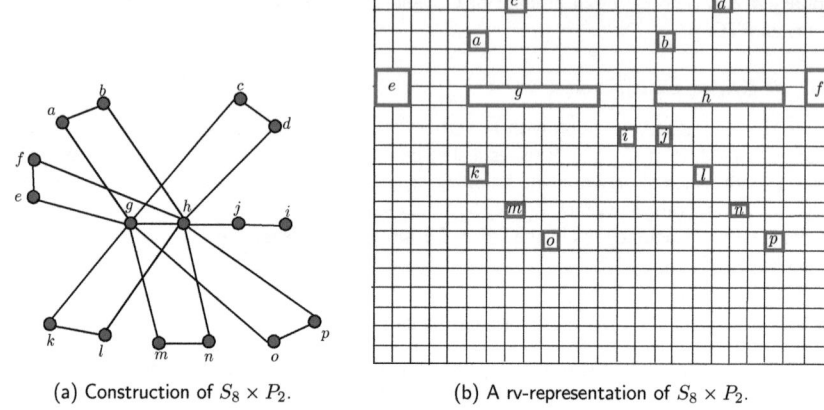

(a) Construction of $S_8 \times P_2$.   (b) A rv-representation of $S_8 \times P_2$.

Figure 4.9: Construction of $S_8 \times P_2$ and its rv-representation.

Similarly, Figure 4.10a represents when $m = 8$, we obtain B_8 $(S_9 \times P_2)$ and its rv-representation is shown in Figure 4.10b. The following result is immediate;

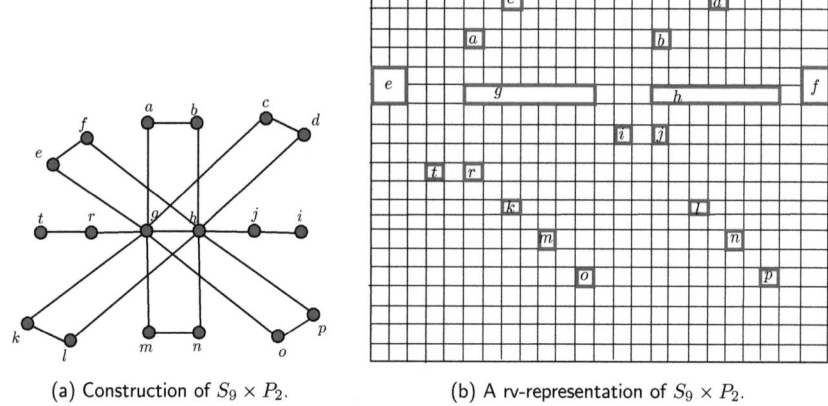

(a) Construction of $S_9 \times P_2$.   (b) A rv-representation of $S_9 \times P_2$.

Figure 4.10: Construction of $S_9 \times P_2$ and its rv-representation.

Corollary 4.2.5. The book graph B_m, for $m = 1, \ldots, 8$, are RVGs.

We can easily construct the RVGs for the cases in Corollary 4.2.5. From Theorem 4.2.1, since the star and path graphs are BVGs, their cartesian products are RVGs in the general case. Next, we investigate cartesian product for complete graphs.

Complete graph: In the cartesian products for complete graphs; $K_2 \times K_2$, $K_3 \times K_2$, $K_3 \times K_3$ and $K_4 \times K_2$ are rectangle-visibility graphs. For $K_3 \times K_4$, $K_4 \times K_4$, $K_4 \times K_5$ and others are not rectangle-visibility graphs since they do not have thickness 2. Hutchinson [3] showed that a rectangle-visibility graph on $n \geq 5$ vertices has at most $6n - 20$ edges. Though, $K_3 \times K_3$ has thickness 3, we have shown in Figure 4.5b that it is an RVG. This agrees with the condition of $6n - 20$ edges for a rectangle-visibility graph since for 9 vertices, we have 34 edges whilst $K_3 \times K_3$ has 18 edges. The construction of $K_4 \times K_2$ and its rv-representation is shown in Figure 4.11.

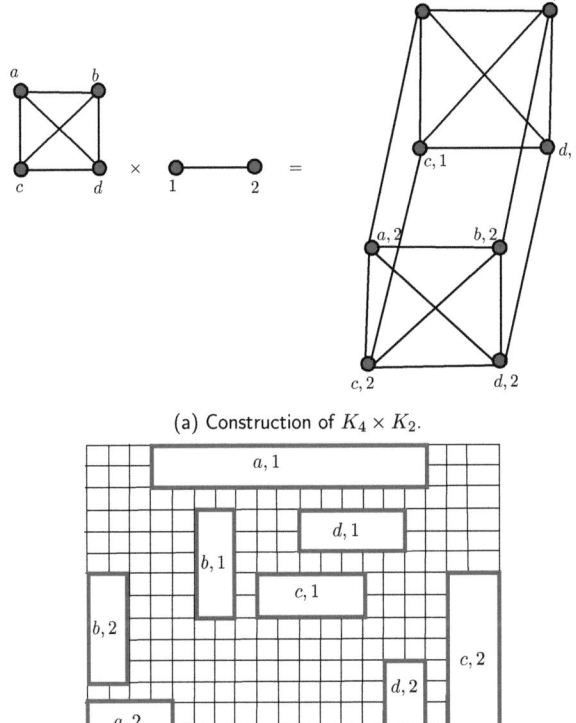

(a) Construction of $K_4 \times K_2$.

(b) A rv-representation of $K_4 \times K_2$.

Figure 4.11: Construction of $K_4 \times K_2$ and its rv-representation.

In the next section, we investigate representations of the direct product for path, cycle and complete graphs as rectangle-visibility graphs.

4.3   Direct Product

Applying the definition given, we investigate representations of the direct product for path, cycle and complete graphs as rectangle-visibility graphs. We start to investigate the direct product for path graphs.

Path graph: We investigate representations of products for $P_3 \square P_3$ and $P_5 \square P_3$ as rectangle-visibility graphs. Figure 4.12 depicts the construction of $P_3 \square P_3$ and its rv-representation. In Figure 4.12a the vertex set, $V(P_3) \times V(P_3) = \{(a,1), (b,1), (c,1), (a,2), (b,2), (c,2), (a,3), (b,3), (c,3)\}$. Without loss of generality, vertex $(a,1)$ is adjacent to vertex $(b,2)$ by the adjacency of a to b in P_3 and also by the adjacency of 1 to 2 in P_3. However, a is not adjacent to c and 1 is not adjacent to 3. $P_3 \square P_3$ is a rectangle-visibility graph which follows a layout in a manner of representing each vertex as a rectangle in a star graph (S_5) layout and a cycle graph (C_4) layout. This is true since $P_3 \square P_3$ consists of two components S_5 and C_4 and is shown in Figure 4.12b.

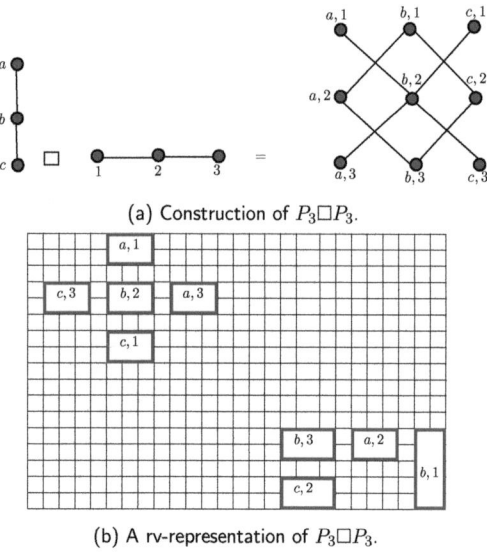

(a) Construction of $P_3 \square P_3$.

(b) A rv-representation of $P_3 \square P_3$.

Figure 4.12: Construction of $P_3 \square P_3$ with its rv-representation.

Also, Figure 4.13 illustrates the construction of $P_5 \square P_3$ with its rv-representation. We follow the same procedure in the definition to obtain the new graph in Figure 4.13a. The rv-representation of $P_5 \square P_3$ is shown in Figure 4.13b.

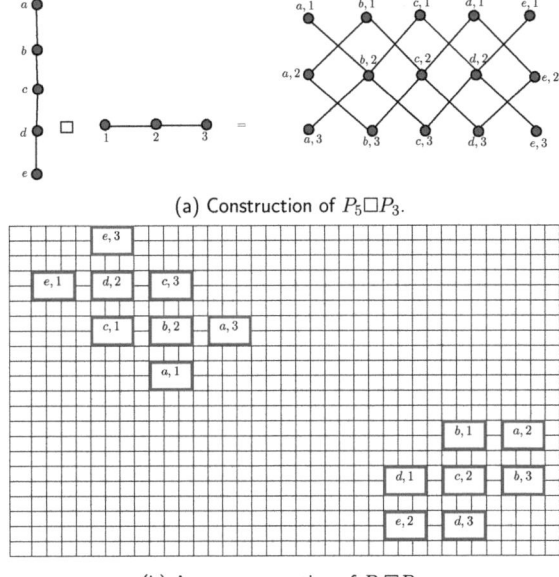

(a) Construction of $P_5 \square P_3$.

(b) A rv-representation of $P_5 \square P_3$.

Figure 4.13: Construction of $P_5 \square P_3$ with its rv-representation.

The direct product of P_n and P_m is a disconnected graph [14]. Hence, we present by construction that $P_n \square P_m$ is a (disconnected) RVG.

Proposition 4.3.1. The direct product of P_n and P_m is a (disconnected) RVG.

Proof. We give the construction.

The direct product of P_n and P_m is a disconnected graph consisting of two components. To represent $P_n \square P_m$ as an RVG, we let each vertex be a rectangle. It is easy to see the visibility for small values of m, n or high values of m with low values of n (and vice versa). For instance, when $m = 2$ for any value of n, each of the components of $P_n \square P_2$ is P_n, which is an RVG. When $m, n \geq 3$, we note that the two components are isomorphic. Each component is asymptotically isomorphic to the grid graph, i.e, the graph possess drawing which can be embedded in $\mathbb{R}^2$ and is planar. We have shown that the grid graph is an RVG hence result. □

Next, we investigate direct products for cycle and path graphs, and finally for complete graphs.

Cycle and path graphs: First we construct the direct product for C_3 and C_3 and next construct $C_6 \square P_2$ and $C_5 \square P_2$. Figure 4.14 depicts the construction of $C_3 \square C_3$ with its rv-representation. In Figure 4.14a, the vertex set $V(C_3) \times V(C_3) = \{(a, 1), (b, 1), (b, 1), (a, 2), (b, 2), (c, 2), (a, 3), (b, 3),$

$(c, 3)\}$ and the edges are formed by the definition given to obtain a new graph. Its rv-representation is shown in Figure 4.14b.

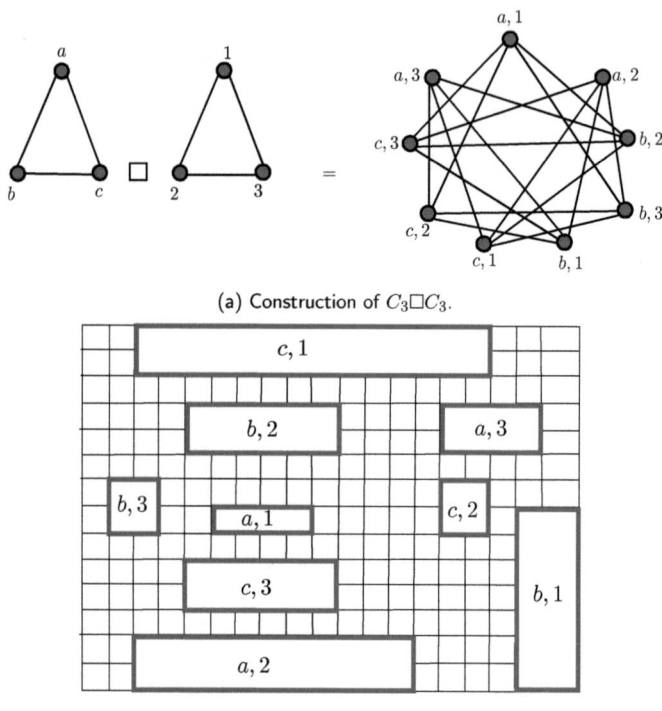

(a) Construction of $C_3 \square C_3$.

(b) A rv-representation of $C_3 \square C_3$.

Figure 4.14: Construction of $C_3 \square C_3$ and its rv-representation.

Now we explore the direct product of cycle and path graphs. Figure 4.15, illustrates the construction of $C_6 \square P_2$ and its rv-representation. The new graph obtained from the construction of $C_6 \square P_2$ is shown in Figure 4.15a while its rv-representation is shown in Figure 4.15b.

Also, the construction of $C_5 \square P_2$ and its rv-representation is shown in Figure 4.16. In the next proposition, we prove by construction that the direct product of C_m and P_2 is an RVG.

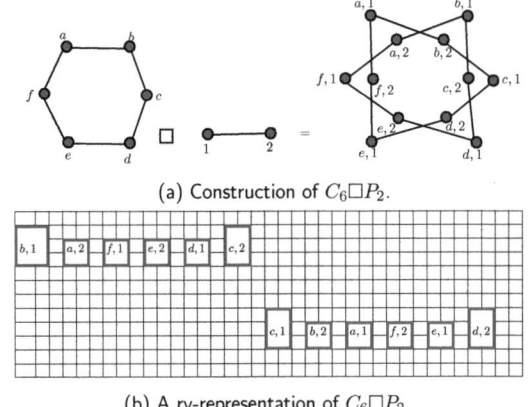

(a) Construction of $C_6 \square P_2$.

(b) A rv-representation of $C_6 \square P_2$.

Figure 4.15: Construction of $C_6 \square P_2$ and its rv-representation.

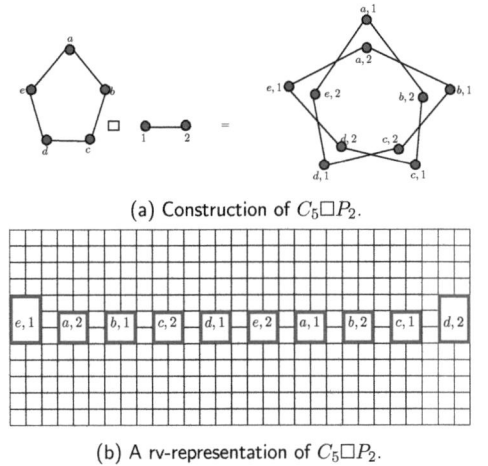

(a) Construction of $C_5 \square P_2$.

(b) A rv-representation of $C_5 \square P_2$.

Figure 4.16: Construction of $C_5 \square P_2$ and its rv-representation.

Proposition 4.3.2. The direct product of C_m and P_2 is an RVG.

Proof. We give the construction.

The direct product of C_m and P_2 produces a new graph that depends on m being odd or even, hence to draw the RVG of $C_m \square P_2$, we consider for odd m and even m separately.

- *Case* 1 *(odd m)*

  For m odd, $C_m \square P_2 = C_{2m}$. Now, draw each vertex as a rectangle following a circular pattern. Place each rectangle depending on the adjacency of the vertices in the graph and continue in that fashion. The last rectangle is visible to the first rectangle. Furthermore, the number of vertices of the cycle graph represents the number of rectangles drawn in the row (horizontal).

- *Case* 2 *(even m)*

  For m even, $C_m \square P_2$ gives a disconnected graph of two components consisting of C_m and C_m and let A and B be the two components. Now, draw each vertex in set A as rectangles depending on the number of vertices of the cycle graph (C_m). Follow the same routine in the odd cycle to obtain the layout in set A. Next, repeat the same procedure in set B and at the end of the layout, we obtain a horizontal visibility between the rectangles.

□

Next, we investigate representations of direct product for complete graph as rectangle-visibility graph.

Complete graph: In the case of the direct product for complete graphs; $K_2 \square K_2$, $K_3 \square K_2$, $K_3 \square K_3$ and $K_4 \square K_2$ are rectangle-visibility graphs. For $K_3 \square K_4$, $K_4 \square K_4$, $K_4 \square K_5$ and others are not rectangle-visibility graphs since they do not have thickness 2. Note that even though, $K_3 \square K_3$ has thickness 3, we have shown in Figure 4.14b that it is an RVG. This agrees with the condition of $6n - 20$ edges for a rectangle-visibility graph since for 9 vertices, we have 34 edges whilst $K_3 \square K_3$ has 18 edges. The construction of $K_4 \square K_2$ and its representation is shown in Figure 4.17.

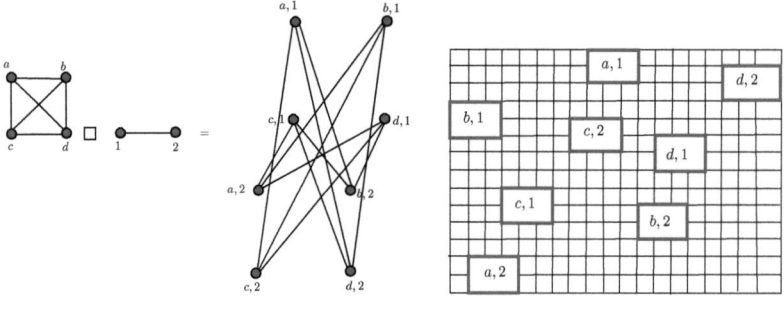

(a) Construction of $K_4 \square K_2$.   (b) A rv-representation of $K_4 \square K_2$.

Figure 4.17: Construction of $K_4 \square K_2$ and its rv-representation.

In the next section, we investigate representations of the strong product for path, cycle and complete graphs as rectangle-visibility graphs.

4.4 Strong Product

Employing the definition given, we investigate representations of the strong product for path, cycle and complete graphs as rectangle-visibility graphs. We begin with strong product for path graphs.

Path graph: We investigate representations for $P_3 \boxtimes P_2$ and $P_3 \boxtimes P_3$ as rectangle-visibility graph. Figure 4.18 depicts the construction of the strong product of P_3 and P_2 and its rv-representation of the vertex set $V(P_3) \times V(P_2) = \{(a,1),(b,1),(c,1),(a,2),(b,2),(c,2)\}$. Without loss of generality, by combining the properties of cartesian and direct products, we obtain a new graph shown in Figure 4.18a as well as its rv-representation in Figure 4.18b.

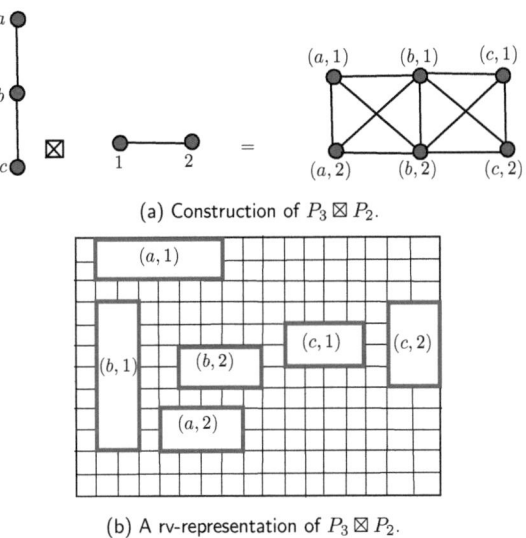

(a) Construction of $P_3 \boxtimes P_2$.

(b) A rv-representation of $P_3 \boxtimes P_2$.

Figure 4.18: Construction of $P_3 \boxtimes P_2$ with its rv-representation.

Next, Figure 4.19 illustrates the construction of strong product of P_3 and P_3 with its rv-representation. We follow the same procedure in the earlier graph to obtain $P_3 \boxtimes P_3$ in Figure 4.19a, and its rv-representation in Figure 4.19b. Following next, is a conjecture on the strong product for path graphs.

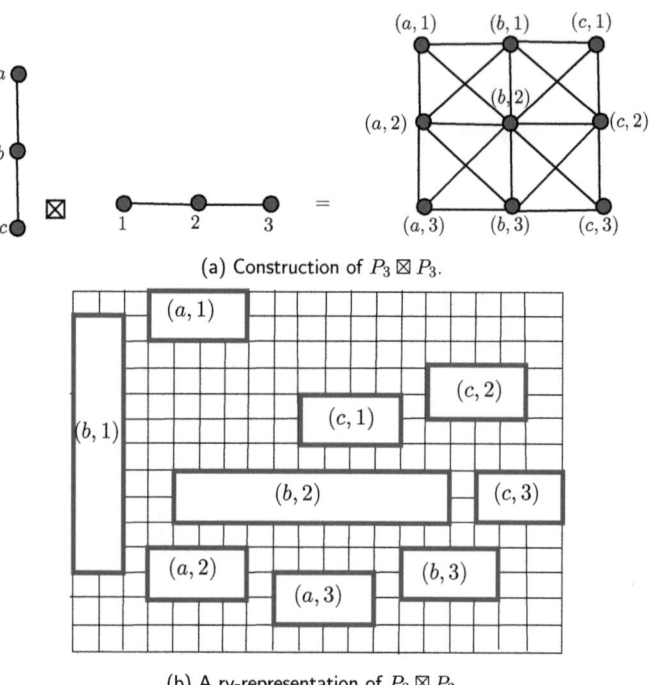

(a) Construction of $P_3 \boxtimes P_3$.

(b) A rv-representation of $P_3 \boxtimes P_3$.

Figure 4.19: Construction of $P_3 \boxtimes P_3$ with its rv-representation.

Conjecture 4.4.1. Among the strong product of the path graphs with n and m vertices, for $1 \leq n, m \leq 4$, $P_n \boxtimes P_m$ and $P_n \boxtimes P_n$ are RVGs.

Next, we investigate the strong product for cycle with path graphs and finally investigate strong product for complete graphs.

Cycle with path graphs: We begin to construct for $C_3 \boxtimes P_2$ and $C_4 \boxtimes P_2$. Figure 4.20, illustrates the construction of $C_3 \boxtimes P_2$ and its rv-representation. The new graph obtained from the construction of $C_3 \boxtimes P_2$ is shown in Figure 4.20a with its rv-representation shown in Figure 4.20b.

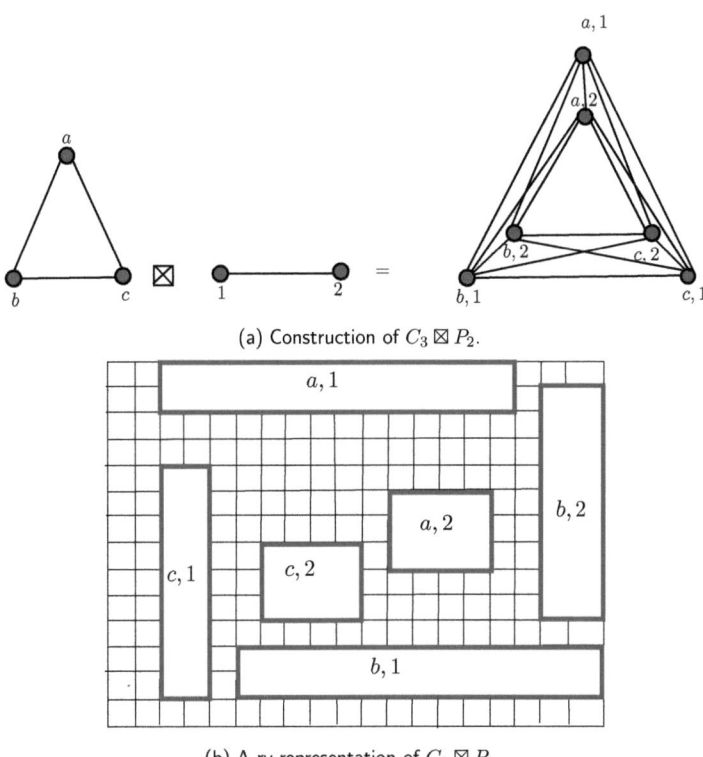

(a) Construction of $C_3 \boxtimes P_2$.

(b) A rv-representation of $C_3 \boxtimes P_2$.

Figure 4.20: Construction of $C_3 \boxtimes P_2$ and its rv-representation.

More so, the construction of $C_4 \boxtimes P_2$ and its rv-representation is shown in Figure 4.21. Next, we give a conjecture on the strong product for cycle and path graphs

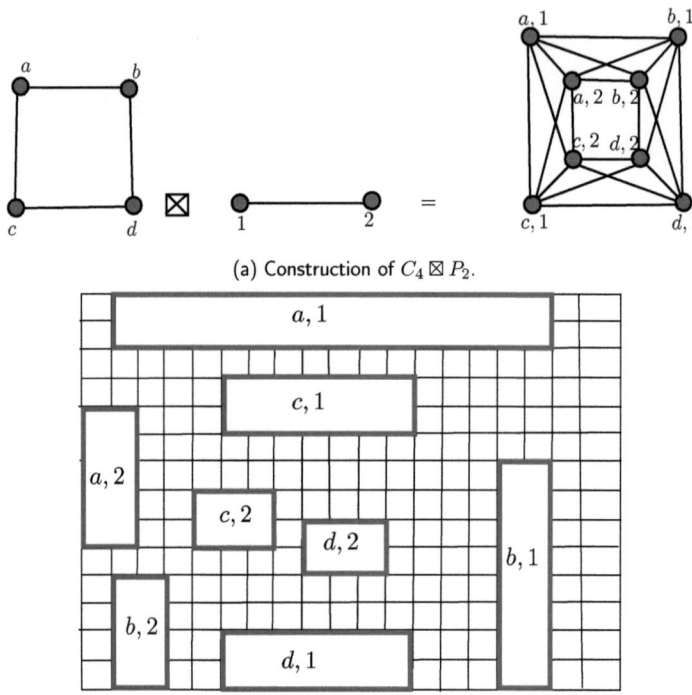

(a) Construction of $C_4 \boxtimes P_2$.

(b) A rv-representation of $C_4 \boxtimes P_2$.

Figure 4.21: Construction of $C_4 \boxtimes P_2$ and its rv-representation.

Conjecture 4.4.2. The strong product of C_m and P_2 for $m = 3, 4, 5$ are RVGs.

Note that the strong product usually introduces more edges making the graphs sometimes no longer RVGs. Finally, we investigate the strong product for complete graphs.

Complete graph: $K_2 \boxtimes K_2$, $K_3 \boxtimes K_2$ and $K_4 \boxtimes K_2$ are rectangle-visibility graphs. For $K_3 \boxtimes K_3$, $K_4 \boxtimes K_3$ and others are not RVGs. In this particular case, $K_3 \boxtimes K_3$ is not an RVG since it does not satisfy the condition $6n - 20$ edges and also has thickness 3. Figure 4.22 shows the construction of $K_4 \boxtimes K_2$ and its rv-representation.

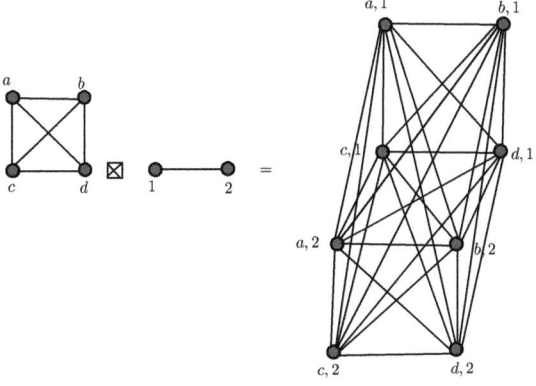

(a) Construction of $K_4 \boxtimes K_2$.

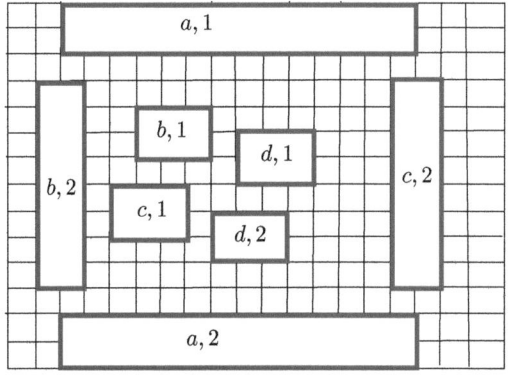

(b) A rv-representation of $K_4 \boxtimes K_2$.

Figure 4.22: Construction of $K_4 \boxtimes K_2$ and its rv-representation.

In this chapter, we have investigated the products of some classes of graphs as rectangle-visibility graphs. We provided constructions for those that are RVGs and also showed why some are not.

5. Conclusion

A rectangle-visibility graph is a graph whose vertices are represented by rectangles and whose edges are represented by vertical or horizontal visibility. There are many types of graph products, however, the three commonly used graph products are cartesian, direct and strong products. The study of graph product is very important in VLSI design.

In this work, we have studied products of fundamental classes of graphs and have classified when their products are rectangle-visibility graphs. In particular, we obtained results on the cartesian, direct and strong products for the path graphs, cycle graphs with path graphs, star graphs with path graphs and complete graphs. We also discussed why the products of some complete graphs are not RVGs.

A possible future study will be to prove the remaining conjectures and investigate bar-visibility representations on products of graphs. This area is ripe with many lines of further investigation.

References

[1] G. Chartand and P. Zhang. *A first course in graph theory*. Courier Corporation, 2012.

[2] A. M. Dean and J. P. Hutchinson. Rectangle-visibility representations of bipartite graphs. *Discrete Applied Mathematics*, 75(1):9–25, 1997.

[3] J. P. Hutchinson, T. Shermer, and A. Vince. On representations of some thickness-two graphs. *Computational Geometry*, 13(3):161–171, 1999.

[4] T. C. Shermer. On rectangle visibility graphs. iii. external visibility and complexity. In *CCCG*, volume 96, pages 234–239, 1996.

[5] E. Peterson. *Rectangle Visibility Numbers of Graphs*. Thesis, Rochester Institute of Technology, 2016.

[6] A. M. Dean, J. A. Ellis-Monaghan, S. Hamilton, and G. Pangborn. Unit rectangle visibility graphs. *The electronic journal of combinatorics*, 15(1):R79, 2008.

[7] M. Garey, D. Johnson, and H. So. An application of graph coloring to printed circuit testing. *IEEE Transactions on circuits and systems*, 23(10):591–599, 1976.

[8] R. Tamassia and I. G. Tollis. A unified approach to visibility representations of planar graphs. *Discrete & Computational Geometry*, 1(1):321–341, 1986.

[9] P. Bose, A. Dean, J. Hutchinson, and T. Shermer. On rectangle visibility graphs. In *International Symposium on Graph Drawing*, pages 25–44. Springer, 1996.

[10] G. Kant, G. Liotta, R. Tamassia, and I. C. Tollis. Area requirement of visibility representations of trees. *Information Processing Letters*, 62(2):81–88, 1997.

[11] A. M. Dean and J. P. Hutchinson. Rectangle-visibility layouts of unions and products of trees. *J. Graph Algorithms Appl*, 2(8):21, 1998.

[12] R. J. Wilson. *An introduction to graph theory*. Pearson Education India, 1970.

[13] V. B. Alekseev and VS. Gončakov. The thickness of an arbitrary complete graph. *Sbornik: Mathematics*, 30(2):187–202, 1976.

[14] R. H. Hammack, W. Imrich, and S. Klavžar. *Handbook of Product Graphs*. CRC press Boca Raton, 2011.

YOUR KNOWLEDGE HAS VALUE

- We will publish your bachelor's and
  master's thesis, essays and papers

- Your own eBook and book -
  sold worldwide in all relevant shops

- Earn money with each sale

Upload your text at www.GRIN.com
and publish for free